J.H. Simpson

The Shortest Route to California

Illustrated by a history of explorations of the great basin of Utah with its

topographical and geological character and some account of the Indian tribes

J.H. Simpson

The Shortest Route to California
*Illustrated by a history of explorations of the great basin of Utah with its
topographical and geological character and some account of the Indian tribes*

ISBN/EAN: 9783743407404

Manufactured in Europe, USA, Canada, Australia, Japa

Cover: Foto ©Andreas Hilbeck / pixelio.de

Manufactured and distributed by brebook publishing software
(www.brebook.com)

J.H. Simpson

The Shortest Route to California

THE

SHORTEST ROUTE TO CALIFORNIA

ILLUSTRATED BY A

HISTORY OF EXPLORATIONS

OF THE

GREAT BASIN · OF UTAH

WITH ITS

TOPOGRAPHICAL AND GEOLOGICAL CHARACTER

AND SOME ACCOUNT OF THE

INDIAN TRIBES.

BY

BREVET BRIG.-GENERAL J. H. SIMPSON, A.M.,

COLONEL CORPS OF ENGINEERS, U. S. ARMY.

PHILADELPHIA:

J. B. LIPPINCOTT & CO.

1869.

PREFACE.

THE writer of the following pages, in the years 1858–59, was Chief Topographical Engineer of the Army of Utah. While serving in this capacity, he explored and opened, under the auspices of the War Department, a wagon route from the valley of Great Salt Lake across the Great Basin of Utah, by which he shortened the distance between Great Salt Lake and San Francisco more than two hundred miles. As the features, topographical, geological, and ethnological, of the country explored by him have never been published, he has deemed it due to the public, the army, and himself, that some *general* account of the same should be presented, and, in connection, a history of the explorations of the Great Basin from the earliest records extant.

A *detailed* report of the writer's explorations in Utah, accompanied by sub-reports on the various subjects connected therewith, from some of the most distinguished professors in the country, was submitted to the Government early in 1861; but its publication has never been ordered by Congress.

(3)

THE GREAT BASIN OF UTAH.

THE country known since the date of the explorations of Frémont in 1843 and '44, by his appellation of the *Great Basin*, has, since the days of Fathers Silvestre Velez Escalante and Francisco Atanacio Dominguez, in 1776, been of great interest. This interest has grown out of its inaccessibility on account of extended deserts; its occupancy by Indians of an exceedingly low type, who subsist chiefly on roots, grass-seed, rats, lizards, grasshoppers, etc.; and the laudable curiosity which prevails in the minds of men to know the physical characteristics of a country which until a very recent period has been a *terra incognita*.

The Great Basin has a triangular shape, nearly that of a right-angle triangle, the mountains to the north of the Humboldt River and Great Salt Lake constituting the northern limit or border, and forming one leg of the triangle; the Sierra Nevada, or western limit, the other equal leg; and the Wasatch Range, to the east of Great Salt Lake, and the short mountain ranges and plateau country to the north of the Santa Fé and Los Angeles Caravan or Spanish trail route, the hypothenuse. These limits are embraced approximately within the 111th and 120th

(5)

degrees of west longitude from Greenwich, and the 34th and 43d degrees of north latitude; or within a limit of nine degrees of longitude and nine of latitude.

The earliest record we have of any examination of this basin is derived from the journal of Father Escalante, descriptive of the travels of himself and party, in 1776–77, from Santa Fé to Lake Utah, by him called *Laguna de Nuestra Señora de la Merced de Timpanogotzis*, and also Lake *Timpanogo;* and thence to Oraybe, one of the villages of the Moquis, and back to Santa Fé. (See accompanying map.) A manuscript copy of this journal, in the Spanish language, is to be found in the rare and valuable library of the late Peter Force, which has recently been purchased by Congress. By this manuscript we learn that Escalante explored as far north, doubtless, as the Timpanogos River, by him called the *Rio San Antonio de Padua;* and as he alludes to the lake, now called Utah Lake, emptying itself into a large body of salt water farther north, there can be no question that he was also cognizant of the existence of the Great Salt Lake.

The destination of Escalante, his journal shows, was Monterey on the Pacific coast; but, being forced most probably by the desert immediately west of Lake Utah to take the so-called southern or Los Angeles route, which Bonneville's party in 1834, and Frémont in 1844, followed, and finding that, while making a great deal of southing, he had made but little progress toward Monterey, his provisions giving out, and fearing the approach of winter, with some difficulty he prevailed upon his party to abandon the idea of reaching Monterey and return to Santa Fé by the way of the villages of the Moquis and of Zuñi.*

The next authentic record which shows that any portion of the Great Basin system was explored at an early date is to be found on the map entitled *"Appendiente al Diario que formo el P. F. Pedro Font del viage que hizo a Monterey y puerto de San Francisco, y del viage que hizo el P. Garces al Moqui. P. F. Petrus Font fecit, Tabutana, Anno 1777,"* which may be freely translated as follows: "A Supplement to the Diary made by Father F. Pedro Font of his journey to Monterey and San Francisco, and of Father Garces' to Moquis, platted by P. F. Petrus Font, at Tabutana, in the year 1777." A copy of this map was sent, several years since, to the Bureau of U. S. Engineers, Washington City, by Captain now Brevet

---

* Humboldt, in his "New Spain," translated by John Black, second edition, London, vol. i. chap. ii p. 22, says, "These regions," referring to those between the Colorado and Lake Timpanogos (Utah Lake), "abounding in rock salt, were examined in 1777 by two travelers full of zeal and intrepidity, monks of the Order of St. Francis, Father Escalante and Father Antonio Velez."

According to the manuscript narrative of these travels by Father Escalante, referred to above, we find that Friar Francisco Atanacio Dominguez, and *not Velez*, was associated with Escalante in these explorations. It is something singular, however, that Escalante's name was Silvestre *Velez* Escalante. Can it be that Humboldt has fallen into the error of making two distinct persons out of this father's name, and of omitting that of Dominguez altogether? Or did a monk by name Antonio Velez explore this region separately from the others and in the same year? We notice, also, that Humboldt dates Escalante's journey A.D. 1777. The manuscript shows that it was commenced July 29, 1776, and terminated in January, 1777.

Major-General Ord, U. S. Army, from an original in the archives of California; and is quite interesting, as showing the large number of Spanish settlements in Middle Sonora at the time of the travels of Fathers Font and Garces, and the exact routes explored by them.

According to this map, Father Garces traveled, as early as 1777 (Humboldt says in 1773[*]), from the mission of San Gabriel, near Los Angeles on the Pacific coast in California, to Oraybe, one of the villages of the Moquis, and his route was along the Rio de las Matires (evidently, from its position, the Mojave). Frémont and others supposed that the Mojave was a tributary of the Colorado and therefore did not belong to the Great Basin system; but this idea was exploded by Lieutenant now Brevet Lieutenant-Colonel Williamson, Major Corps of U. S. Engineers, in 1853, and afterward by Lieutenant now Brevet Major-General Parke, Major Corps of Engineers, in 1855; both of whom fully determined that this stream sank, and that between it and the Colorado was a ridge which separated the waters.[†]

In this connection, it may be interesting to observe that Humboldt, speaking of the delay on the part of the Spaniards, notwithstanding their enterprising spirit, in opening communications between New Mexico and California, holds the following language:

"The letter post still (at the date of his researches, in 1803–4) goes from this post (San Diego) along the

---

[*] See his "New Spain," vol. ii. p. 268.
[†] Pacific Railroad Reports, vol. v. p. 33, and vol. vii. p. 3.

northwest coast to San Francisco. This last estab-
lishment, the most northern of all the Spanish posses-
sions of the new continent, is almost under the same
parallel with the small town of Taos in New Mexico.
It is not more than 300 leagues distant from it; and
though Father Escalante, in his apostolical excur-
sions in 1777, advanced along the western bank of the
river Zagnananas toward the mountains *de los Gua-
caros*, no traveler has yet come from New Mexico to
the coast of New California. This fact must appear remarkable to those who know, from the history of the
conquest of America, the spirit of enterprise and the
wonderful courage with which the Spaniards were ani-
mated in the sixteenth century. Hernan Cortez landed
for the first time on the coast of Mexico in the district
of Chalchinhcuecan in 1519, and in the space of four
years had already constructed vessels on the coast of
the South Sea at Zacatula and Tehuantepec. In 1537,
Alvar Nuñez Cabeza de Vaca appeared with two of
his companions, worn out with fatigue, naked, and
covered with wounds, on the coast of Culiacan, opposite the peninsula of California. He had landed with
Panfilo Narvaez in Florida, and after two years' ex-
cursions, wandering over all Louisiana and the northern part of Mexico, he arrived at the shore of the
great ocean in Sonora. This space, which Nuñez went
over, is almost as great as that of the route followed
by Captain Lewis from the banks of the Mississippi
to Nootka and the mouth of the river Columbia.*

---

* "This wonderful journey of Captain Lewis was undertaken
under the auspices of Mr. Jefferson, who by this important ser-

When we consider the bold undertakings of the first Spanish conquerors in Mexico, Peru, and on the Amazons River, we are astonished to find that for *two centuries the same nation could not find a road by land in New Spain from Taos to the port of Monterey.*"*

Humboldt here was undoubtedly in error. The map of Father Font, before referred to, shows that as early as 1777 Father Garces traveled from the mission of San Gabriel, near the Pacific coast, to Oraybe, one of the villages of the Moquis, in New Mexico. And the Spanish inscriptions found by Lieutenant J. H. Simpson, Corps of Topographical Engineers, U. S. Army, on the rock (*El Moro*) near Zuñi in New Mexico, an account and fac-similes of which he gives in his "Journal of a Military Reconnoissance from Santa Fé to the Navajo Country in 1849,"† show that there was as early as 1716 a communication opened with the Moquis from Santa Fé. The inscription, translated, is as follows: "In the year 1716, upon the 26th day of August, passed by this place Don Felix Martinez, Governor and Captain-General of this kingdom, for the purpose of reducing and uniting Moqui" (a couple of words here not deciphered). The manuscript of Father Escalante's journal, before referred to, also shows that there was a well-known road from Oraybe, *via* Zuñi, to Santa Fé, and which his party followed. These facts show that at least

vice rendered to science has added new claims on the gratitude of the *savans* of all nations." (Note by Humboldt.)

* Humboldt's "New Spain," vol. ii. pp. 289, 290.

† See Sen. Doc. 64, 31st Cong., 1st Sess., p. 123; or same published by J. B. Lippincott & Co., Philadelphia, p. 104.

as early as 1777, and most probably as early as 1773 (the date, according to Humboldt, of Garces' journey to Oraybe), there was a communication all the way from Santa Fé, and, without doubt, from Taos, *via* Moqui, to San Gabriel; and, as Father Font's map shows, even all the way to Monterey and the Bay of San Francisco.

Greenhow, in his "Oregon and California,"* represents that "in 1775 Friars Font and Garces traveled from Mexico, through Sonora and the country of the Colorado River, to the mission of San Gabriel in California, making observations on their way with a view to the increase of intercourse between Mexico and the establishments in the latter region. They were, however, coldly received by their brethren, who informed them that they had no desire to have such communications opened; and their journal was never made public." Their map, which Mr. Greenhow seems, however, not to have seen, shows that Father Font traveled from the Presidio of San Miguel, situated in about lat. 29° 30′ and long. 110° 30′ (probably on the Rabasaqui), as far as the port of San Francisco; and that Garces traveled only from San Gabriel to Moqui. It also shows that the "Rio de San Philipe," on some old maps, was, in all probability, Kern River.

Greenhow remarks that the journals of the expeditions of Friars Escalante, Garces, and Font are still preserved in manuscript in Mexico; but, "from all accounts, are of no value." · Humboldt, on the contrary,† speaks highly of the information imparted by

---

* 2d edition, p. 114.   † "New Spain," vol. ii. p. 253.

Font and Garces; and so far as the journal of Escalante is disparaged by Greenhow, we are convinced his criticism is unjust; for not only is this journal written in a very plain, unpretending, direct way, but it abounds in excellent and apparently just observations; and it is wonderful that the courses and distances should plat so correctly, and should agree so well with our present maps.

The next published account we find of any portion of the Great Basin country is in the memoir of Lieutenant now Brevet Major-General Gouverneur K. Warren, Major Corps of Engineers, U.S.A. In this memoir is a letter to General Warren from Robert Campbell, Esq., a well-known gentleman of Saint Louis, who was long connected with the fur trade and its operations in the tramontane regions of the West.*   In this letter Mr. Campbell gives *verbatim* the statement of Mr. James Bridger, corroborated by Mr. Samuel Tolleck, both Indian traders, to the effect that he, Bridger, was the first discoverer of Great Salt Lake, in the winters of 1824 and 1825.

Mr. Bridger further states, in Mr. Campbell's letter, that "in the spring of 1826 four men went in *skin* boats around it (the Great Salt Lake), to discover if any streams containing beaver were to be found emptying into it, but returned with indifferent success." Washington Irving, in his "Bonneville's Adventures," revised edition, page 186, says, "*Captain Sublette*, in one of his early expeditions across the

---

* Lieutenant Warren's Memoir, vol. xi.; Pac. R. R. Reports, p. 35.

mountains, is said to have sent four men, in a *skin* canoe, to explore the lake, who professed to have navigated all round it; but to have suffered excessively from thirst, the water of the lake being extremely salt, and there being no fresh streams running into it.

"Captain Bonneville doubts this report, or that the men accomplished the circumnavigation; because, he says, the lake receives several large streams from the mountains which bound it to the east."

It would thus appear that Sublette in all probability was the person who sent out the four men referred to by Bridger, in a *skin* canoe to explore the lake; and, though Bonneville doubts the report of the occurrence, yet the testimony of Bridger is corroborative of the fact; and the circumstance of its being an actual fact that there are no fresh-water streams coming into the lake on its west shore, along its whole length, certainly accounts for the thirst of Sublette's party. It may be true that Sublette's party did not discover the fresh-water streams running into the lake from the south and east; but this only shows that they did not explore the lake thoroughly; not that they did not explore it at all.

The next authentic account of any discoveries within the Great Basin, we find given in "Bonneville's Adventures," by Washington Irving. Colonel Bonneville, U. S. Army, it would appear, was the first explorer to cross, in 1832, the Rocky Mountains into the valley of Green River, *with wagons.**

---

* "Captain Bonneville now considered himself as having fairly passed the crest of the Rocky Mountains; and felt some degree

To quote from Irving: "On the 24th July, 1833, by his (Captain Bonneville's) orders, a brigade of forty men set out from Green River Valley to explore the Great Salt Lake. They were to make the complete circuit of it, trapping on all the streams which should fall in their way, and to keep journals and make charts, calculated to impart a knowledge of the lake and the surrounding country. All the resources of Captain Bonneville had been tasked to fit out this favorite expedition. The country lying to the southwest of the mountains, and ranging down to California, was as yet almost unknown; being out of the buffalo range, it was untraversed by the trapper, who preferred those parts of the wilderness where the roaming herds of that species of animal gave him comparatively an abundant and luxurious life. Still it was said the deer, the elk, and the big-horn were to be found there; so that, with a little diligence and economy, there was no danger of lacking food. As a precaution, however, the party halted on Bear River and hunted for a few days, until they had laid in a supply of dried buffalo meat and venison; they then passed by the head-waters of the Cassie River, and soon found themselves launched on an immense sandy desert. Southwardly, on their left,

---

of exultation in being the first individual that had crossed, north of the settled provinces of Mexico, from the waters of the Atlantic to those of the Pacific, *with wagons*. Mr. William Sublette, the enterprising leader of the Rocky Mountain Fur Company, had two or three years previously reached the valley of the Wind River, which lies on the northeast of the mountains; but had proceeded with them no further." (Bonneville's Adventures, rev. ed., p. 61.)

they beheld the Great Salt Lake, spread out like a sea; but they found no stream running into it. A desert extended around them, and stretched to the southwest, as far as the eye could reach, rivaling the deserts of Asia and Africa in sterility. There was neither tree, nor herbage, nor spring, nor pool, nor running stream; nothing but parched wastes of sand, where horse and rider were in danger of perishing.

"Their sufferings at length became so great that they abandoned their intended course and made toward a range of snowy mountains, brightening in the north, where they hoped to find water. After a time, they came upon a small stream leading directly toward these mountains. Having quenched their burning thirst, and refreshed themselves and their weary horses for a time, they kept along this stream, which gradually increased in size, being fed by numerous brooks. After approaching the mountains, it took a sweep toward the southwest, and the travelers still kept along it, trapping beaver as they went, on the flesh of which they subsisted for the present, husbanding their dried meat for future necessities.

"The stream on which they had thus fallen is called by some Mary's River, but is more generally known as Ogden's River, from Mr. Peter Ogden, an enterprising and intrepid leader of the Hudson's Bay Company, who first explored it."*

---

* It must be borne in mind that the Humboldt River constitutes a portion of the Great Basin system. Lieutenant Warren, in his Memoir, p. 36, says, "The party from the Hudson Bay Company, referred to in the postscript to Mr. Campbell's letter, was under the enterprising leader, Mr. Peter Ogden, who dis-

"The trappers continued down Ogden's River, until they ascertained that it lost itself in a great swampy lake, to which there was no apparent discharge. They then struck directly westward, across the great chain of California mountains intervening between these interior plains and the shores of the Pacific.*

"For three and twenty days they were entangled among these mountains, the peaks and ridges of which are in many places covered with perpetual snow. Their passes and defiles present the wildest scenery, partaking of the sublime rather than the beautiful, and abounding with frightful precipices. The sufferings of the travelers among these savage mountains

---

covered the Ogden's or Mary's River in 1828. One of Mr. Ogden's party took a woman for his wife from among the Indians found on this river, to whom the name of Mary was given. From this circumstance the stream came to be called Mary's River. It is also called Ogden's River, after its discoverer."

Lieutenant Warren might with more propriety, we think, have said that the stream formerly was called Ogden's or St. Mary's River; but since the explorations of Frémont in 1845–46 it has been known, by emigrants and others, entirely as the *Humboldt* River, the name Frémont gave it.

* Irving is here in error. Walker did not go directly *westward* from the swamp (sink) of the Ogden's River (the Humboldt) across the great chain of California mountains (the Sierra Nevada); but, striking southwardly, continued down along their *east* side, for nearly five degrees of latitude, before he crossed them near their southern termination, by a pass since known as *Walker's Pass.* We get this information from Mr. E. M. Kern, the assistant of Frémont, who ten years subsequently was guided by Walker over this very route.

were extreme; for a part of the time they were nearly starved; at length they made their way through them, and came down upon the plains of New California, a fertile region extending along the coast, with magnificent forests, verdant savannas, and prairies that look like stately parks. Here they found deer and other game in abundance, and indemnified themselves for past famine. They now turned toward the south, and passing numerous small bands of natives, posted upon various streams, arrived at the Spanish village and post of Monterey."*

It would thus seem that Walker and his party failed in exploring around the west portion of the Great Salt Lake, on account of the desert in that region, and were forced to take a route along the northern section of the Great Basin to California; and it is represented by Irving that on their return they turned the Sierra Nevada at its southern extremity. This being the case, it is likely they took the Spanish trail route, which Frémont, ten years after, in 1844, followed, and on which, at Vegas de Santa Clara, he overtook this same Joseph Walker, in charge of a trading party.

The next authentic account we have of any explorations of the Great Basin is from the report by Colonel Frémont of his expedition, in 1843 and 1844, to Oregon and California, through the South Pass, when, on the 6th of September of the former year, he attained the summit of a butte near the mouth of the Weber River, whence he saw for the first time

---

* "Bonneville's Adventures," pp. 326–328.

2

the waters of Great Salt Lake.* Forming an encampment near the mouth of the Weber, he remained in the vicinity a few days, to make some observations and take a hasty sketch of the lake.

Subsequently, in continuation of his expedition, he explored in the following winter from Fort Vancouver along the east base of the Sierra Nevada, or along what may be called the northwestern edge of the Great Basin, as far as the vicinity of Johnson's Pass, where he crossed the Sierra to the valley of the Sacramento. On his return east, in the spring of 1844, he turned the Sierra Nevada at its southern extremity, got upon the Spanish trail along the Mojave River in the Great Basin, crossed the Rio Virgen and other tributaries of the Colorado, and near Las Vegas de Santa Clara again entered the Great Basin, and explored it along its south and eastern edge up to the eastern portion of Lake Utah, where he left it and crossed the dividing ridge into the valley of Green River.

Colonel Frémont's report shows that at the time of this expedition he had not seen the previously published history and map of the explorations of Bonneville, for had he done so he would probably not have been led into the error to which he attributed most of his hardships, of constantly looking for the hypothetical river of Buenaventura, which, as he supposed, took its rise in the Rocky Mountains and emptied into the Bay of San Francisco, and upon which he ex-

---

* Frémont's Report, Cong. Doc. No. 166, p. 151, published in 1845.

pected to winter. His language is as follows: "In our journey across the desert, Mary's Lake, and the famous Buenaventura River, were two points on which I relied to recruit the animals and repose the party. Forming, agreeably to the best maps in my possession, a connected water-line from the Rocky Mountains to the Pacific Ocean, I felt no other anxiety than to pass safely across the intervening desert to the banks of the Buenaventura, where, in the softer climate of a more southern latitude, our horses might find grass to sustain them, and ourselves be sheltered from the rigors of winter and from the inhospitable desert."*

Touching this question, Colonel Bonneville, in a letter to Lieutenant Warren on the subject of his explorations in and west of the Rocky Mountains, uses the following language; and, as it bears upon the fact as to whom should be accorded the credit of the discovery of the Great Basin, we think proper to make an extract from it:†

"GILA RIVER, N. M., August 24, 1857.

"DEAR SIR:—I thank you for your desire to do me justice as regards my map and explorations in the Rocky Mountains. I started for the mountains in 1832. * * * * * I left the mountains in July, 1836, and reached Fort Leavenworth, Missouri, the 6th of August following. During all this time I kept good account of the course and distances, with

* Frémont's Report for 1843–44, p. 205. See also pp. 196, 214, 219, 221, 226, 255.

† Lieutenant Warren's Memoir, vol. xi.; Pac. R.R. Reports.

occasional observations with my quadrant and Dolland's reflecting telescope. * * * * * I plotted my work, found it proved, and made it into three parts: one a map of the waters running east to the Missouri State line; a second, of the mountain region itself; and a third, which appears to be the one you have sent me, of the waters running west. On the map you send, I recognize my names of rivers, of Indian tribes, observations, Mary's or Maria's River, running southwest, ending in a long chain of flat lakes, never before on any map, and the record of the battle between my party and the Indians, when twenty-five were killed. This party clambered over the California range, were lost in it for twenty days, and entered the open locality to the west,. not far from Monterey, where they wintered. In the spring they went south from Monterey, and turned the southern point of the California range to enter the Great Western Basin. On all the maps of those days the Great Salt Lake had two great outlets to the Pacific Ocean: one of these was the Buenaventura River, which was supposed to head there;* the name

---

* Colonel Bonneville is here probably in error. On Finley's Map of North America (Philadelphia, 1826), given by Lieutenant Warren in his Memoir, p. 30, and which purports to include all " the recent geographical discoveries" up to the date stated, the Buenaventura is represented not as one of the outlets of Great Salt Lake into the Pacific, but as the outlet of *Lake Salado*, doubtless Lake Sevier of our present maps. The two rivers which are represented on this map as disemboguing from the Great Salt Lake into the Pacific are the Rio Los Mongos and Rio Timpanogos. The fact of Father Escalante, in 1776, giving the name of *Buenaventura* to a river, evidently, from the plotting

of the other I do not recollect. It was from my explorations and those of my party alone that it was ascertained that this lake had no outlet; that the California range *basined* all the waters of its eastern slope without further outlet; that the Buenaventura and all other California streams drained only the western slope. It was for this reason that Mr. W. Irving named the Salt Lake after me, and he believed I was fairly entitled to it.

* * * * * * * *

" Yours, etc.,
"B. L. E. BONNEVILLE,
" *Colonel 3d Infantry.*
" LIEUTENANT G. K. WARREN,
" *Topographical Engineers.*"

It must nevertheless be stated, in justice to Frémont, that though Bonneville's map ignores the Buenaventura River, and all the others which on the old maps had been represented as flowing from the Basin into the Pacific; yet that this fact and that of the existence of the Great Basin, which Frémont has so well brought out in his report, have not been descanted on at all by Irving; and thus Frémont may very naturally not have been impressed with the discoveries which Colonel Bonneville has more recently brought out significantly in his letter to Lieutenant Warren.

---

of his notes, Green River, and which he supposed flowed westwardly from the Rocky Mountains into Lake Salado (Sevier), the limits of which have been left undetermined on Humboldt's map, points, we think, to the origin of the Rio Buenaventura, and its subsequent hypothetical extension from Lake Sevier to the Bay of San Francisco.

We would also state that it would appear from Colonel Frémont's report that it was a favorite purpose of his, on his return from California, to *cross* the Great Basin *directly*, instead of turning it at its southern extremity. He is speaking of what occurred as he was turning the southern end of the Sierra Nevada, by the Tah-e-chay-pah Pass, to get on the Spanish trail: "In the evening a Christian Indian rode into the camp, well dressed, with long spurs, and a *sombrero*, and speaking Spanish fluently. It was an unexpected apparition, and a strange and pleasant sight in this desolate gorge of a mountain— an Indian face, Spanish costume, jingling spurs, and horse equipped after the Spanish manner. He informed me that he belonged to one of the Spanish missions to the south, distant two or three days' ride, and that he had obtained from the priests leave to spend a few days with his relations in the Sierra. Having seen us enter the *pass*, he had come down to visit us. He appeared familiarly acquainted with the country, and gave me definite and clear information in regard to the desert region east of the mountains. I had entered the pass with a strong disposition to vary my route, and to travel directly across toward the Great Salt Lake, in the view of obtaining some acquaintance with the interior of the Great Basin, while pursuing a direct course for the frontier; but his representation, which described it as an arid and barren desert, that had repulsed by its sterility all the attempts of the Indians to penetrate it, determined me for the present to relinquish the plan; and, agreeably to his advice, after crossing the Sierra, continue our intended route along its eastern base to the Spanish trail."*

Thus, like Father Escalante and Walker, Frémont was foiled from directly crossing the Great Basin on account of its reported arid nature; and evaded it by keeping along its southern edge.

The next authentic account we have of any explorations within the Great Basin is to be found in the pamphlet entitled "Geographical Memoir upon Upper California, in illustration of his map of Oregon and California, by John Charles Frémont, addressed to the Senate of the United States."† This Memoir and the accompanying map show that Colonel Frémont, in his explorations of 1845, entered the Great Basin by way of the Timpanogos River;‡ followed down the valley of Utah Lake, and its outlet, the Jordan River, to its mouth in Great Salt Lake; turned this lake at its southern borders; passed westwardly by

---

* Frémont's Report, p. 254.

† Senate Miscellaneous Doc., No. 148, 3d Cong., 1st Session.

‡ Frémont's map represents, and Lieutenant Warren's Memoir gives currency to the error, that Frémont passed from the Duchesne's Fork, *up Morin's Fork,* and thence across the divide to the Timpanogos. This is a physical impossibility; for Morin's Fork, or White Clay Creek, as it is now called, is a tributary of the Weber, and instead of running into Duchesne's Fork, and being thus a tributary of the Colorado, is, on the contrary, a tributary of the Great Salt Lake In other words, Duchesne's Fork and Morin's Fork are on opposite sides of the divide,— the Uinta range. The mistake on Frémont's map has arisen evidently from carelessness on the part of the draughtsman; and Lieutenant Warren, knowing nothing personally of the error, naturally has given currency to it.

Pilot's Peak to Whitton's Spring; and thence his party was divided,—Mr. E. M. Kern, with Joseph Walker as guide, striking northwestwardly for the Humboldt (Mary's) River, following it down to its sink,* and striking southwardly to and passing along the east shore of Carson Lake, to Walker's River; and Colonel Frémont, with Carson and Godey as guides, and a portion of the party, striking south-westwardly more directly across the Great Basin to Walker's Lake, where the parties again met. Here separating again, Mr. Kern, guided by Walker, proceeded southwardly to the head of, and along, Owen's River and lake, and thence to Walker's Pass of. the Sierra Nevada, where he left the Basin and crossed the Sierra into the valley of Lake Tulare and the Rio San Joaquin. Frémont, on the contrary, traveled northwardly to Carson River, where he crossed it at the same point as in his preceding exploration; and thence to Salmon Trout Creek, up which he traveled, and crossed the Sierra Nevada in latitude 39° 17′ 12″ N., or 38·2 miles north of his pass of 1844.

Lieutenant Warren, in his Memoir,† has erroneously reversed the respective positions of Frémont and Kern in their explorations after separating at Whitton's

----

* It is generally understood that Frémont was the first to establish the wagon route along the Humboldt. This is a mistake.

The route Kern's party took down the Humboldt was already a well-beaten wagon route, which had been used by Hastings and other California emigrants for several years previously to the explorations of Frémont.

† Page 50.

Spring. He makes Frémont to explore the route along the Humboldt River and Carson Lake, whereas Kern was the topographer of the party which explored *that* route, with Walker as guide; and Frémont explored the route more directly across the Basin to Walker's Lake, which Warren as erroneously has attributed to Kern and Walker.

The Geographical Memoir of Frémont, as already stated, does not enter into the particulars of his exploration of 1845 and 1846; but only gives a general view of the Great Basin. This view is graphic, and, in the main, so far as the present writer's observations have extended, is just, and corrects some errors into which, from imperfect data, he had fallen, in his previous explorations. The idea which he had entertained of the Basin's being made up of a system of small lakes and rivers scattered over a flat country, was found to be entirely untrue, and, on the contrary, he found that the *mountain* structure predominated. The long stretch of mountain range, however, which on his map is represented as being the continuation westwardly of the Wasatch Range, and as separating the waters of the Great Basin from those of the Colorado, is evidently hypothetical, and has not been corroborated by subsequent explorers. This view, however, in no way militates against the theory and fact of the Great Basin system as one distinct from the valley of the Colorado; because, as is to be seen in many instances in the Basin itself, a very slight rim or rise of ground may be the divide between distinct sub-basin systems.

The next authentic account, in the order of dates,

which we have of explorations within the Great Basin,
is to be found in the report by Captain Howard Stansbury, Topographical Engineers, of his "Exploration
and Survey of the Valley of the Great Salt Lake
of Utah, in 1849," published by order of Congress.
This report, however erroneous it may have been
in its discussions of the Mormon question at that
early date, and however its conclusions may have
been falsified by the history of this people since the
date of the report, we cannot but regard, in a geographical and physical point of view, as of great
value.   We have had occasion, in many instances in
our reconnoissances west of the Rocky Mountains, and
in the region of the Great Salt Lake, to test the accuracy
of Captain Stansbury's work; and it has been a grati-
fication to us to find that his report and map have
represented the country so correctly and been of so
much service to us.   To him and his assistant, the
lamented Captain Gunnison, Topographical Engineers,
the public is indebted for a thorough triangular survey
of the Great Salt Lake; and to them is the credit due
of a complete exploration of the lake around its en-
tire limits; a feat which Joseph Walker, by Colonel
Bonneville's directions, attempted, as before stated,
sixteen years previously, but which, on account of
the desert lying on its west, and the consequent want
of fresh water, he failed to execute.   Stansbury, however, extended his explorations in the Great Basin
only as far as Pilot Knob, a prominent landmark,
sixty-four miles due west from Great Salt Lake.

The next authentic account of explorations in the
Great Basin is that by Captain E. G. Beckwith, 3d

Artillery, the assistant of Captain Gunnison, Topographical Engineers, in his expedition for the survey of a railroad route, near the forty-first parallel of latitude, who took charge of the expedition after the massacre of Gunnison and a portion of his party by Indians on Sevier River, on the 26th October, 1853. The party entered the Great Basin from the valley of Green River, by the Wasatch Pass, and a creek he calls Salt Creek, a branch of the Sevier;* and thence they returned to the usually traveled *southern* route from Los Angeles, and proceeded by the way of Nephi, Payson, Provo, etc., to Great Salt Lake. In the ensuing year, 1854, Captain Beckwith explored some of the tributaries of Great Salt Lake and Utah Lake, issuing from the Wasatch and Uinta Mountains, and, passing over the southern end of Great Salt Lake, he struck generally a north of west course, across the Great Basin, to the Humboldt Pass of the Humboldt Range; thence southwardly, in Ruby Valley, to the Hastings Road Pass of this same range; and thence,

---

* Messrs. Beale and Heap passed over nearly this same route in 1853, in advance of Captain Gunnison's party, and, after reaching Vegas de Santa Clara, took the Spanish trail route to California. (See Heap's Journal, published by Lippincott, Grambo & Co., 1854.) Colonel Frémont also subsequently, during the winter of 1853–54, followed very nearly the route of Captain Gunnison to Grand River, and thence to Parowan and Cedar City, on the Spanish trail; thence his course was directly west over the Great Basin to the Sierra Nevada, which, on account of snow, he was obliged to cross by Walker's Pass, some sixty to eighty miles to the southward. (Frémont's letter to the editor of the *National Intelligencer* of June 13, 1854. House Mis. Doc. No. 8, 2d Sess., 33d Cong.)

northwestwardly across the mountains lying south of the Humboldt, to Lassen's Meadows, on the Humboldt River. Thence his course was westwardly, through the valley of the Mud Lakes, to the Madelin Pass of the east range of the Sierra Nevada, where he left the Great Basin.* It will be noticed that, *up to that time*, this was the most direct exploration which had been made across the Great Basin from Great Salt Lake City; but yet it *was too far north and too tortuous to be of great value as affording a direct wagon route to Placerville, Sacramento, and San Francisco.* Besides, as a wagon route to Lassen's Meadow, we believe it has never been used.

The next report we have of an attempt being made to cross the Great Basin directly from Great Salt Lake City, toward Walker's Lake, for the purpose of *avoiding the great detour by the Humboldt River and getting the shortest route to San Francisco,* is to be found in the report of Captain Rufus Ingalls to the Quartermaster-General, dated August 25, 1855; giving an account of the movements of Colonel Steptoe's command to, and from, Great Salt Lake City in 1854 and '55. His language on this point is as follows:

"The wagon routes across the continent are so very rough in mountainous regions, and always quite circuitous, particularly from Great Salt Lake City to the Bay of San Francisco, that Colonel Steptoe took measures to have the country lying directly west explored for a more nearly air-line road. Two Mormons were engaged as principal explorers, and di-

---

* Pacific R. R. Report, vol. ii.

rected to explore from the south end of the Great Salt Lake, on the Beckwith route, or near to it, to Carson Valley. This party left the lake in September, and returned the following November. It proved quite an extensive trip, owing, in my present opinion, to the tricky character of the Mormons. They made a most flattering report. They said they had discovered a wagon road, along which a command could move with ease, etc., saving one hundred and fifty or two hundred miles. The colonel had not seen Lieutenant Beckwith's report, nor had he any other information than that given by his exploring party; but, being deeply sensible of the importance to the Territory of Utah and the overland immigrants of laying out and opening a more direct and practicable road than the crooked ones now traveled, he determined to take his command and the large wagon train over this new route.

"As spring approached, however, the chief Mormon, who had agreed to act as guide, became rather restive, and evinced an unwillingness to go, which caused the colonel to distrust him, and shook his confidence in the report he had made of the road. As a matter of security, another party was organized under 'Porter Rockwell,' a Mormon, but a man of strong mind and independent spirit, a capital guide and fearless prairie-man. He went out as far as the great desert tracts lying southwest of the lake, and very nearly on a level with it, and found that at *that season* they could not be passed over, unless with wings, and returned. It proved fortunate that we did not undertake the march with O. B. Huntington as guide.

The march would have been disastrous; though Rockwell and others are of the opinion that by going on a line some thirty miles farther south, along the foot of mountains seen in that direction, a fine road can be laid out, avoiding, in a great degree, the desert. I believe such to be the case myself. I am clearly of the opinion that a suitable officer could, by a proper reconnoissance, lay out a road passing by 'Rush Valley,' turning southwest, and going by New River, Walker's Lake, into Carson Valley, and save two hundred miles' distance. This route having been declared impracticable, the colonel decided to pass around the north end of the lake, and thence by the Humboldt to Carson Valley."*

It thus seems that Colonel Steptoe was deterred from attempting a direct route across the Great Basin toward San Francisco, by the reports which he had received, and took the old roundabout road by way of the Humboldt River.

We have now, as we believe, exhausted the subject of the Explorations in and around the Great Basin, up to the time when the writer reported for duty as chief engineer with the army under General Albert Sidney Johnston, in Utah, in August, 1858. This history shows that up to this period a direct road toward San Francisco, from Great Salt Lake, or Camp Floyd, across the Great Basin, had never been thor-

---

* Appendix A, Qr.-Master-Gen.'s Report, accompanying Sec. War's Annual Report, 1855, constituting Ex. Doc. No. 1, House of Rep., p. 156, 34th Cong., 1st Session.

oughly attempted; but that in every instance, from fear of encountering reported deserts, explorers had shrunk from the task. It was universally believed in Utah that at this period not even a Mormon had ventured to cross the Basin in this direct manner toward Carson or Walker's Lake. Some, more adventurous than others, had made a less circuitous bend than the old route by the Humboldt River; but not one had accomplished a *direct* journey across.

It was the failure on the part of others to accomplish this desirable exploration, as well as the possible advantages of a *new and short road,* that stimulated the writer to make, through General Johnston, a *project* of exploration to the War Department, which had in view the accomplishment of this very enterprise; and thus, if possible, the opening of a wagon road that would be of benefit to the army and to the nation.

His project of exploration was approved by General Johnston; sanctioned by the War Department; and under the authority of the latter, the expedition was ordered, and consequently received the complete outfit it did from the general commanding. The party consisted of Captain J. H. Simpson, Topographical Engineers (now Colonel of Engineers and Brevet Brigadier-General U. S. Army), in command of expedition; assistants Lieutenant J. L. Kirby Smith, Topographical Engineers, U. S. Army, in charge of astronomical observations with sextant for latitude and time or longitude; Lieutenant Haldeman L. Putman,* Topographical Engineers, U. S.

---

* Lieutenant Smith was mortally wounded while "changing front forward" with his regiment to repulse a desperate attack of

Army, in charge of compass survey of route and topography of country, observations with astronomical transit for longitude, and with dip circle and magnetometer for magnetic dip, declination, and intensity; Henry Engelmann, Geologist, Meteorologist, and Botanical Collector; Charles S. McCarthy, Collector of Specimens of Natural History, and Taxidermist; C. C. Mills, Photographer; Edward Jagiello and Wm. Lee, assistants to Astronomer, Meteorologist, and Photographer; and Mr. Reese as Guide, and "Pete," a Ute Indian, assistant.

The party had twelve six-mule wagons to carry supplies, and two spring-wagons to convey the instruments; and, with the Topographical and Quartermaster's employés and military escort, all told, numbered sixty-four men. Lieutenant Alexander Murry, 10th Infantry, U. S. Army, commanded the escort, and Assistant-Surgeon Joseph C. Baily, U. S. Army, accompanied the expedition as surgeon. The party was rationed for three months, and starting from Camp Floyd (since called Camp Crittenden), in Utah Lake Valley, forty miles south of Great Salt Lake City, on the 2d of May, 1859, it successfully crossed the Great Basin in a general course south of

---

the rebels on Battery Robinett, in the battle of Corinth, October 3d and 4th, 1862. For "gallant and meritorious services" in this battle, he was brevetted colonel.

Lieutenant Putman was killed, July 18, 1863, at the head of his troops in their assault of Fort Wagner, S. C., and brevetted same day for "gallant and meritorious" services.

Both Smith and Putman were brave and accomplished Union officers, and the service lost none more promising during the war.

west to Genoa, near the head of Carson River at the east foot of the Sierra Nevada, and returned to Camp Floyd on the 5th of August following; thus accomplishing the reconnoissance in three months and three days, the provisions lasting "to a notch," and without the loss of a single man or horse.

The result of the expedition was the opening of two new, practicable wagon routes across the Great Basin; the shorter of which lessened the distance between Great Salt Lake City and San Francisco a trifle over two hundred miles; and the other about one hundred and eighty miles. Immediately the first-mentioned became the postal route; the "Pony Express" commenced its trips over it, and emigrants to California have used it ever since. Also by the recommendation of Captain Simpson, and the efforts of Colonel Bee, the then President of the Overland Telegraph Company, which at that date had extended its wires only from San Francisco a distance of two hundred and fifty miles to Genoa, Congress was induced to pass the bill incorporating the Overland Telegraph Company and authorizing it to construct a telegraph across the continent from the Atlantic to the Pacific. It was the fortunate circumstance of Captain Simpson finding so feasible a telegraph route and reporting it to Colonel Bee, that induced the latter to go on to Washington from California and press the matter of the Overland Telegraph through Congress to a successful result.

The report of this expedition of Captain Simpson cost him no inconsiderable labor; and, illustrated as it is by a complete map, meteorological profiles, and numerous sketches, and supported by reports on the geological, botanical, and meteorological character of the country traversed, from the most distinguished scientific men of the country, and giving information of a region over which it is believed no white man ever traveled before the expedition referred to; it is to be regretted that Congress has not yet ordered the report, as repeatedly recommended by the Engineer Department, to be published. Twice have the Committees on Printing in the Senate reported favorably as to the character of the report and expediency of its publication; but either from the expense, or some other cause, the results of the expedition, though costing the government at least sixty thousand dollars, are to be found only in the Bureau of Engineers at Washington; and thus the emigrants who make use of the road, and the Pacific Railroad Companies who have been constantly asking for it for their purposes, have been deprived of all the benefit of the explorations. It is to be hoped that some member of Congress who has an eye to the interests of the country at large, and who is not willing that all the cost and labor of the expedition referred to, over so interesting a section of country, should be lost, will yet ask for the publication of this report and have influence enough to get it ordered by Congress. Assuredly, the expense of printing the map, profiles, and text of the report would not be much; and this Congress should at least do, if it finds that to include the sketches would make the cost of the publication too great.

Having said thus much with regard to the important results of the author's expedition, it may not prove uninteresting to the reader to be informed of some of the chief characteristics of the country explored by him.

The first thing which will strike one, on looking at the map, is the great number of mountain ranges which the routes cross in the Great Basin. This will appear the more remarkable, as the idea has been generally entertained, since the explorations of Frémont in 1843 and 1844 (though, as before remarked, he corrected the error on his succeeding expedition), that this Great Basin was a *flat country scattered over with a system of small lakes and rivers,* and destitute of mountains. The fact, on the contrary, is that it is probably the most mountainous region, considering its extent, within the limits of our country, and so far from being scattered over with a system of small lakes and rivers, which seems to imply a considerable number of this kind of water area, it has but a limited number of lakes, and they almost entirely confined to the bases of the great Sierras which bound the Basin.

These lakes are—proceeding from north to south, and around the circumference of the Great Basin— Great Salt Lake, Lake Utah, Sevier Lake, and Small Salt Lake, on the eastern side of the Basin; and on the west, proceeding from south to north, Owen's Lake, Mono Lake, Walker's Lake, the two Carson Lakes, Humboldt Lake, Pyramid Lake, the Mud Lakes, and Lake Abert. Besides these, there are Franklin Lake and Goshoot Lake, to the east of the East Humboldt Range. These constitute all the lakes that have been discovered in the Great Basin, and they are all without outlet. Great Salt Lake is seventy miles long and from twenty to thirty broad; Pyramid and Walker's Lakes, the next largest, are both about thirty miles long by ten wide; all the others are smaller. Pyramid Lake, Walker's Lake, and Utah Lake, which are all fresh-water lakes, abound in fine large trout.

The principal rivers which, on account of their width and depth, require bridging or ferry in their flush state, during the time of melting snow, are the Bear, Weber, Roseaux or Malade, Jordan, Timpanogos, Spanish Fork, and Sevier Rivers, which have their sources in the Wasatch Mountains, on the east side of the Basin, and flow into lakes near the base of these mountains; the Mojave, Owen's, Walker's, Carson, and Truckee or Salmon Trout, which have their sources in the Sierra Nevada, and flow into lakes at their base and sink; and the Humboldt River, which flows from east to south of west along the northern portion of the Basin and sinks. The longest of these is the Humboldt, about three hundred miles long, and the next longest Bear River, about two hundred and fifty miles long. The others vary from forty to one hundred and twenty miles in length. In width they vary from about fifty to one hundred and fifty feet, and in depth from two to fifteen feet, depending upon the season and locality.

All the other streams are of small extent, and, taking their rise in the many mountain ranges by which the Basin is traversed, generally from north to south, they seldom flow beyond their bases, where in the alluvion they sink. These streams are usually so small that one can jump across them, and seldom require bridging. The large as well as the small streams mentioned, when not brackish, not unfrequently contain trout. One of these small streams is *Reese* River, called so by Captain Simpson after his chief guide. This river has since become famous on account of the rich silver-bearing rocks with which its valley is characterized, and its being also the site of the city of Austin, which so suddenly sprang into existence after the discovery of the precious metal.

The trend of the mountain ranges is almost invariably north and south, the limits of variation being between the true and the magnetic north. The mountains rise quite abruptly from the plains, and form bases varying in breadth from a few miles to about twelve. These mountain ranges are so frequent and close together as to make the areas between them more like valleys than plains. In cross section the valleys are slightly concave; and Captain Simpson in his survey crossed them, in a direction of south of west, on the average every ten or fifteen miles. In length they are commensurate with the mountain ranges. Longitudinally, or in a general direction north and south, they are nearly level.

The most massive and lofty mountains, commencing at Camp Floyd and proceeding westward, are the O-quirr, Guyot, Goshoot or Tots-arr, Un-go-we-ah, Mon-tim, Humboldt, We-ah-bah, Pe-er-re-ah, and Se-day-e ranges. Of these the Tots-arr, Un-go-we-ah, Humboldt, Pe-er-re-ah, and Se-day-e are the most massive and lofty. The lengths of the ranges in some instances were at least one hundred and twenty miles, and they then extended into unknown regions beyond the field of Captain Simpson's explorations. These ranges attain in the case of Union Peak (so called by Captain Simpson), the highest point of the Tots-arr or Goshoot Range, an altitude above the plain of from five thousand to six thousand feet, or of from ten thousand to eleven thousand feet above the sea. In the case of the O-quirr Range, the highest point (Camp Floyd Peak), according to Lieutenant Putman's measurement, by theodolite, was found to be four thousand two hundred and fourteen feet above the camp at its foot; and as this locality, by barometric measurement, is four thousand eight hundred and sixty feet above the sea, the peak referred to is nine thousand and seventy-four feet above the sea. The *highest pass* was on Captain Simpson's return route, and through the Un-go-we-ha Range. By barometric measurement it was eight thousand one hundred and forty feet above the sea. The passes are all, with but little difficulty, surmountable by wagons; but their barometrical profiles show that they are too steep for railroad purposes. These barometrical profiles of Captain Simpson, to which the Union Pacific Railroad Company have had access, have already been of very material service in obviating the great expense of another survey, to which the company would otherwise have been obliged to resort.

The chief agricultural characteristic of the country traversed is desert, the exceptions being as follows: On Captain Simpson's more northern route, in the case of the *large valleys between the mountain ranges* and going westward from Camp Floyd—Rush Valley, Pleasant Valley (the valley of Fish or Deep Creek, not on the route but in the vicinity of Pleasant Valley), Ruby Valley, Walker's Valley, and Carson Valley. All these are cultivable in limited portions. And on his return route, going eastward from Genoa, Carson Valley (common to outward route), Steptoe Valley, Antelope Valley, and Crosman Valley. The altitude of these valleys above the sea varies from three thousand eight hundred and forty feet, the lowest depression of Carson Valley, to six thousand one hundred and forty-six feet, the altitude of Steptoe Valley. Carson Valley has already shown its capacity to grow the small cereals and garden vegetables; and we doubt not the other valleys named, though higher in altitude, will be found sufficiently warm to mature the growth of the more hardy cereals and plants. Captain Simpson's return or more southern route, though about thirty miles longer, is much the best in respect to cultivable valleys and grass.

The other exceptions to the desert character of the Basin are the *small narrow valleys and ravines* of the mountain streams, which, taking their rise high up in the mountains, course down to the plains or main valleys and sink. These valleys, though rich, are generally too high above the sea, and therefore too cold, for arable purposes; but are valuable as furnishing in great abundance the small mountain bunch-grass, which has fattening qualities almost if not quite equal to those of oats.

Another exception to the universal characteristic of desert is *the abundance of the dwarf cedar*, which is to be seen on almost every one of the mountain ridges, and which high up in the mountains is not unfrequently intermingled with the *pine* and *mountain mahogany*. The abundance of this cedar, as well as occasional supply of other kinds of timber, has made Captain Simpson's routes, independent of their being the shortest across the Great Basin, decidedly the most practicable for the overland telegraph.

The portion of the country traversed which may be called *unqualifiedly desert* is, on his *more northern* route, the region between Simpson's Springs in the Champlin Mountains, and the Sulphur Springs at the east base of the Tots-arr or Goshoot Range, a distance of eighty miles; albeit the grass and water at Fish Springs intervene, to make the greatest distance between water and grass forty-eight and a half miles; between the west base of the Se-day-e Mountains and Carson Lake, a distance of fifty miles; and between Carson Lake and Walker's River, a distance of twenty-one miles. On Captain Simpson's *return* or *more southern route*, between Carson River and Carson Lake, a distance of twenty-three miles; and between the Perry range and the Champlin Mountains, a distance of one hundred and three miles; though Chapin's Springs and Tyler Spring, with their limited pasture-ground, and the Good Indian Spring, with its small supply of water but abundance of grass, within this interval alleviate in a very material degree this last stretch and take it out of the category of *continuously* unmitigated desert.

In relation to the propriety of the term, Great Basin, being applied to this region of country, we remark, that if by it the idea is conveyed that this great area is chiefly one of a hydrographic character,—that is, filled with lakes and rivers,—it is so far a misnomer. Erroneous also is the idea that because it is called a *Basin* it must, as a whole, present a generally *concave* surface. The truth is, it is only a Basin inasmuch as the few lakes and streams that are found within it *sink* and *have no outlet to the sea.*

It may also be considered as made up of several *minor or subsidiary basins;* and, regarding them in succession, not in the order of magnitude, we have—

1st. Lake Sevier Basin. Elevation of lowest point above the sea, slightly less than 4690 feet.

2d. Great Salt Lake Basin. Elevation of lowest point above the sea, 4170 feet.

3d. Humboldt River Basin. Elevation of lowest point above the sea, near (Beckworth) Lassen's Meadows, 4147 feet.

4th. Carson River Basin. Elevation of lowest point above the sea, at Carson Lake, 3840 feet.

5th. Walker's River Basin. Elevation of lowest point above the sea, seven miles above Walker's Lake, 4072 feet.

(Walker's Lake Basin, estimated at about same as Carson), 3840 feet.

6th. Owen's Lake Basin. Altitude unknown.

7th. Mojave River Basin. Elevation of lowest point above the sea (Williamson), 1111 feet.

All these valleys or basins, it will be noticed, are on the *outskirts* of the Great Basin, just within its circumference; and as the valleys of the *great central area* have an average altitude of about five thousand five hundred feet, which is for much the larger portion of the area about fifteen hundred feet higher than said basin, and for the Mojave portion over four thousand feet higher, it will at once be apparent that, as a whole, the Basin should be conceived as an elevated central region extended over much the greater portion, and, in proximity to the circumference, sloping toward the sub-basins bordering the circumference. When this idea is entertained, and this extended *central portion* is in addition conceived of as being traversed by high and extensive ranges of mountains, on an average about fifteen miles apart, ranging north and south and forming intermediate valleys of commensurate lengths; bearing in mind at the same time that the order of depression of the basins is from Lake Sevier, where it is least, around successively by Great Salt Lake, Humboldt River Valley, Carson Lake, Walker's Lake, to the valley of the Mojave, where it is much the greatest; a very good mental daguerreotype can be had of the Great Basin inside of its inclosing mountains. From this description we think it will be obvious that while the so-called Great Basin is in some small degree a Basin of lakes and streams, *it is pre-eminently a Basin of mountains and valleys!*

In regard to the geological character of the mountains within the Great Basin, Captain Simpson's explorations show that from Camp Floyd west, as far as about Kobeh Valley, those of carboniferous origin predominate; though over the desert proper, between Simpson's Springs and the Tots-arr Range, the igneous are a characteristic, and near the Humboldt Range those of the Devonian age obtain. From Kobeh Valley to the Sierra Nevada the ranges are almost exclusively of igneous origin, and present few indications of stratified rocks. The knowledge, geologically, of this extensive *terra incognita,* for the first time given to the government in the reports of Captain Simpson's assistant, Mr. Engelmann, and by Mr. Meek, the palæontologist, is an interesting result of the expedition, and goes far to fill up the gap that remained to complete the geological profile of our country from the Atlantic to the Pacific, on the line of Captain Simpson's explorations. These reports not only discuss the geology and palæontology of the Great Basin, but also *of the whole route through from Fort Leavenworth to the Sierra Nevada;* and to no two geologists probably could the work have been better assigned, since Mr. Engelmann, independent of his scientific and practical ability, was the geologist of Lieutenant Bryan's expedition to the Rocky Mountains in 1856, and of Captain Simpson's expedition, from Fort Leavenworth to the Sierra Nevada and back, in 1858 and 1859; and Mr. Meek's well-earned reputation certainly pointed him out as the most capable person to whom to refer the palæontological discoveries of the expedition. In this connection it may be also proper to state that Mr. Engelmann, in his sub-reports, has devoted a great deal of space to the discussion of the *meteorological phenomena* of the Great Basin, and, illustrating as he does his views by accompanying diagrams, his report will prove of great value to science in this particular.

With regard to the *Indians* of the Great Basin,
Dr. Garland Hurt, the intelligent and brave Indian
agent in Utah during the Mormon difficulty in 1857,
1858, and 1859, and the only civil officer connected
with the general government whom the Mormons
could not drive out of their Territory, has furnished
Captain Simpson with a very interesting memoir.
From this memoir it appears that the Indians of the
Great Basin, including those of the valleys of Green
and Grand Rivers, consist of two tribes; the *Ute* and
the *Sho-sho-nes* or *Snakes.*

The *Ute* tribe Dr. Hurt divides into the *Pah-Utahs,*
Tamp-Pah-Utes, Chevериches, Pah-Vants, San-Pitches,
and *Py-èdes.*

The *Utahs* proper inhabit the waters of Green
River, south of Green River Mountains, the Grand
River and its tributaries, and as far south as the Navajo country. They also claim the country bordering on Utah Lake, and as far south as the Sevier
Lake. They are a brave race, and subsist principally by hunting. The buffalo having left their
country and gone east over the Rocky Mountains,
their hunting this game in the country of the Arra-
pahoes and Cheyennes brings them in continual con-
flict with those tribes. Dr. Hurt says it is his opinion,
from a familiar acquaintance with them, that there is
not a braver tribe to be found among the aborigines of
America than the Utahs, none warmer in their at-
tachments, less relenting in their hatred, or more capable of treachery. Their chief in 1859 was *Arra-
pene*, the successor of the renowned *Wacca*, sometimes
erroneously called Walker. Some of the superior

bands, both of the Snakes and Utahs, are nearly always in a state of starvation, and are compelled to resort to small animals, roots, grass-seed, and insects for subsistence. The general government has opened farms for these Indians in the valleys of the Spanish Fork and San Pete.

The *Pah-vants* occupy the Corn Creek, Paravan, and Beaver Valleys, and the valley of Sevier. On Corn Creek they have a farm under the supervision of the general government. It was a portion of this tribe that is reported to have massacred Captain Gunnison and a number of his party in 1858; though Mr. J. Forney, Superintendent of Indians in Utah, in his report of September 29, 1859, fixes the stigma of this horrible outrage on the Mormons.

The *Py-edes* live adjoining the Pah-vants, down to the Santa Clara, and are represented as the most timid and dejected of all the Utah bands. They barter their children to the Utes proper for a few trinkets or bits of clothing, by whom they are again sold to the Navajos for blankets, etc. They indulge in a rude kind of agriculture, which they probably derived from the old Spanish Jesuits. Their productions are corn, beans, and squashes. The Mountain Meadow massacre is ascribed by the Mormons to them; but, as Dr. Hurt justly remarks, "any one at all acquainted with them must perceive at once how utterly absurd and, impossible it is for such a report to be true."

The *Sho·sho·nes* Dr. Hurt divides into Snakes, Bannacks, To-si-witches, Go-sha-utes, and Cum-um-pahs, though he afterward classes the last two divisions as hybrid races between the Sho-sho-nes and the Utahs.*

The *Snakes* are fierce and warlike in their habits, and inhabit the country bordering on Snake River, Bear River, Green River, and as far east as Wind River. They are well supplied with horses and fire-arms, and subsist principally by hunting. They are the enemies of the Crows and Blackfeet, on account of the buffalo having disappeared from their country west of the Rocky Mountains, and their being obliged to hunt them as trespassers on the ter-ritory of these tribes east of the mountains. They have also been at war with the Utes for several gen-erations. They, however, profess friendship for the

---

* Mr. J. Forney, Superintendent of Indian Affairs in Utah, classes and numbers the various tribes and bands of Indians in Utah as follows:

" Sho-sho-nes, or Snakes	4,500
Bannacks	500
Uinta Utes	1,000
Spanish Fork and San Pete farms	900
Pah-vant (Utes)	700
Pey-utes (South)	2,200
Pey-utes (West)	6,000
Elk Mountain Utes	2,000
Washoe of Honey Lake	700
	18,500

" The Sho-sho-nes claim the northeastern portion of the ter-ritory for about four hundred miles west and from one hundred to one hundred and twenty-five miles south from the Oregon line. The Utes claim the balance of the territory." (Pres. Mes. and Doc., 1859–60, Part I.)

whites; and it is their boast that under their chief, Wash-i-kee, the blood of the white man has never stained their soil. It is certain, nevertheless, that small parties of this band, living in Box Elder county, with some Bannack Indians from Oregon, robbed, during the season of 1859, three parties of emigrants on the emigration roads to the north and east of Great Salt Lake, and killed ten or twelve of their number.

The *Bannacks* inhabit the southern borders of Oregon, along the old Humboldt River emigrant road, and have the reputation of infesting that portion of the route, and of being of a very thievish, treacherous character.

The *To-sa-witches,* or *White Knives,* inhabit the region along the Humboldt River, and, according to Dr. Hurt, have the reputation of being very treacherous; though we believe they have proved quite friendly of late years. Captain Simpson met them ranging in small parties between the Un-go-we-ah Range and Cooper's Range on his more southern route.

The *Go-shoots* Dr. Hurt classes among the Sho-shones; but according to Mr. George W. Bean, Captain Simpson's guide in the fall of 1858, who has lived in Utah ever since the Mormons entered this region, and has been frequently employed as interpreter among the Indians, they are the offspring of a disaffected portion of the Ute tribe that left their nation, about two generations ago, under their leader or chief, Go-ship, whence their name Go-ship-utes, since contracted into Go-shutes. Captain Simpson is disposed to believe that they are thus derived, from the fact that he noticed among them several Utes, who, while claiming that they belonged to the Utes proper, had intermarried with the Go-shoots and were living among them.

These Go-shoots are few in number, not more, probably, than two or three hundred, and reside principally in the grassy valleys west of Great Salt Lake, along and in the vicinity of Captain Simpson's routes, as far as the Un-go-we-ah Range.

In addition to the Indians just mentioned as inhabiting the Great Basin, should be mentioned the *Py-ute* and the *Washoe* tribes, which, not being within Dr. Hurt's jurisdiction, were not included by him.

The Py-utes, according to Major Dodge, their Indian agent in 1859, numbered at that date between six thousand and seven thousand souls. They inhabit Western Utah, from Oregon to New Mexico; their locations being generally in the vicinity of the principal rivers and lakes of the Great Basin, viz., Humboldt, Carson, Walker, Truckee, Owen's, Pyramid, and Mono. They resemble in appearance, manners, and customs the Delawares on our Missouri frontier, and with judicious management, and assistance from government, would in three years equal them in agriculture. Their chief in 1859 was Won-a-muc-ca (the Giver), and it was a portion of this tribe, under this chief, who had been engaged just previously in the massacres in Western Utah. Their language resembles in some words the Sho-sho-ne, yet it differs so much from it that Captain Simpson's guide, Ute Pete, who spoke both Ute and Sho-sho-ne, could not understand them. This tribe is frequently confounded with the Pah-utes, with which they show only a distant affinity.

The *Washoes*, according to Major Dodge, numbered in 1859 about nine hundred souls, and inhabit the country along the eastern slope of the Sierra Nevada, from Honey Lake, on the north, to the Clara, the west branch of Walker's River, a distance of one hundred and fifty miles. They are not inclined to agricultural pursuits, nor any other advancement toward civilization. They are destitute of all the necessaries to make life even desirable. In 1859 there was not one horse, pony, or mule in the nation. They are peaceable, but indolent. In the summer they wander around the shores of Lake Bigler, in the Sierra Nevada, principally subsisting on fish. In the winter they lie around in the *artemisia* (wild sage) of their different localities, subsisting on a little grass-seed. The Indian vocabulary appended to Captain Simpson's report shows that they are a distinct tribe, and in no way assimilate with the Utes, Sho-sho-nes, or Py-utes.

The Indians all along Captain Simpson's routes, from Great Salt Lake to Carson River, are of the very lowest type of mankind, and forcibly illustrate the truth which the great physicist of our country, Professor Arnold Guyot, of the College of New Jersey, has brought out so significantly in his admirable work, "The Earth and Man," to wit, *that the contour, relief, and relative position of the crust of the earth are intimately connected with the development of man.* These Indians live in a barren, and in winter, on account of its altitude, a cold, climate; and the consequence is that they are obliged to live entirely on rabbits, rats, lizards, snakes, insects, rushes, roots, grass-seed, etc. They are more filthy than beasts, and live in habitations which, summer and winter, are nothing more than circular inclosures, about four feet high, without roof, made of the *artemisia* or sage bush, or branches of the cedar, thrown around on the circumference of a circle, and which serve only to break off the wind. As the temperature in the winter must at times be as low as zero, and there must fall a good deal of snow, it will readily be perceived that they must suffer considerably. Anything like a covered lodge, or *wick-e-up* of any sort, to protect them from the rain, cold, or snow, Captain Simpson did not see among them. Their dress, summer and winter, is a rabbit-skin tunic or cape, which comes down to just below the knee; and seldom have they leggins or moccasins. The children at the breast were perfectly naked, and this at a time when overcoats were required by Captain Simpson's party. The women frequently appeared naked down to the waist, and seemed unconscious of any immodesty in thus exposing themselves.

The fear of capture causes these people to live some distance from the water, which they bring in a sort of jug made of willow tightly platted together and smeared with fir-gum. They also make their bowls and seed and root baskets in the same way; a species of manufacture quite common among all the Indian tribes, and which Captain Simpson saw in his Explorations of 1849, in the greatest perfection, among the Navajos and Pueblo Indians of New Mexico.

Captain Simpson describes, in his report, a visit to one of their *kants*, as they call their habitations, as follows :

"Just at sunset, I walked out with Mr. Faust to see some of these Go-shoots at home. We found, about one and a half miles from camp, one of their habitations, which consisted only of some cedar branches disposed around the periphery of a circle about ten feet in diameter, and in such a manner as to break off, to the height of about four feet, the wind from the prevailing direction. In this inclosure were a number of men, women, and children. Rabbit-skins were the clothing generally; the poor infant at the breast having nothing on it. In the center was a brass kettle, suspended to a three-legged crotch or tripod. In this they were boiling the meat we had given them. An old woman superintended the cooking, and at the same time was engaged in dressing an antelope-skin. When the soup was done, the fingers of each of the inmates were stuck into the pot and sucked. While this was going on, an Indian, entirely naked with the exception of his breech-cloth, came in from his day's hunt. His largest game was the rat, of which he had quite a number stuck around under the girdle about his waist. These he threw down, and they were soon put by the old woman on the fire and the hair scorched. This done, she rubbed off the crisped hair with a pine knot, and then, thrusting her finger into the paunch of the animal, pulled out the entrails. From these pressing out the offal, she threw the animals, entrails and all, without further cleaning, into the pot."

Mr. Reese, Captain Simpson's guide, avers that he has seen them roast their rats without in any way cleaning them, and then eat them with great relish.

The rats are caught by a dead-fall, made of a heavy stone and supported by a kind of figure 4. They are also speared in their holes by a stick turned up slightly at the end and pointed; and with another of spade-form at the end, the earth is dug away until the animal is reached and taken.

The Go-shoots, as well as the Diggers, constantly carry about with them these instruments, which, with the bow and arrow and net, constitute their chief means for the capture of game. The nets, made of excellent twine fabricated of a species of flax which grows in certain localities in this region, are three feet wide and of very considerable length. With this kind of net they catch the rabbit, as follows. A fence or barrier made of the wild-sage bush plucked up by the roots, or cedar-branches, is laid across the paths of the rabbits, and on this fence the net is hung vertically. The rabbits are then driven from their lairs, and, in running along their usual paths, are intercepted by the net and caught in its meshes.

The only large game they have is the antelope, and this they are seldom able to kill. Their mode of taking him is as follows. They make a sort of trap inclosure of a V-shape, formed by two fences of indefinite lengths, composed of cedar-branches, and converging from a wide open mouth to a point. Within the inclosure and near the vertex of the angle a hole is dug, and in this the Indian secretes himself with his bow and arrow. The antelope, being driven into the mouth of the trap, is naturally directed by the fence on either side to make his escape at the angle. Reaching this point, the Indian, whom he has just passed, pops up from his hiding-place and shoots him.

Their mode of starting a fire is certainly very primitive, and is described in Captain Simpson's journal of June 3d, as follows:

"On reaching our camping-place, which I call the Middle Gate, I saw a naked Indian stretched out on the rocks on an inclination of about twenty degrees. He was so much the color of the rocks, that he escaped our notice till we were right upon him. On being aroused, he looked a little astonished to see so many armed white men about him, but soon felt assured of his safety by our kind treatment. He seemed particularly pleased when he saw the long string of white-topped wagons coming in, and laughed outright for joy. I counted twenty-seven rats and one lizard lying about him, which he had killed for food. He had with him his appliances for making fire. They consisted simply of a piece of hard 'grease-wood' (so called) about two feet long, and of the size or smaller than one's little finger, in cross-section. This was rounded at the butt. Then a second flat piece of the same kind of wood, six inches long by one broad and one-half thick. This second piece had a number of semispherical cavities on one face of it. With this laid on the ground, the cavities uppermost, he placed the other stick between the palms of his hands, and with one end of the latter in the cavity, and holding the stick in a vertical position, he would roll it rapidly

forward and back till the friction would cause the
tinder, which he had placed against the foot of the
stick in the cavity, to ignite. In this way I saw him
produce fire in a few seconds."

As illustrative of the character of these Indians,
and the kind of country to which they attach the
most value, the writer gives one more extract from
his journal of May 27:

"An old Digger Indian has visited our camp, and
represents that we are the first white persons he has
ever seen. He says there are a large number of Indians
living around, but they have run away from fear of
us. I asked him why he had not been afraid. He
said he was so old, that it was of no consequence if
he did die. I told him to say to them that we would
be always glad to see them, and whenever they saw
a white man, always to approach him in a friendly
way, and they would not be hurt. He has been round
eating at the different messes, and at length had so
gorged himself as to be unable to eat more, until he
had disgorged, when he went around again to renew
the pleasure.

"I showed him my watch, the works of which he
looked upon with a great deal of wonder. He said
he would believe what I told him about the magnetic
telegraph, the next time he was told it. He is at
least sixty years old, and says he has never had a
chief. I asked him if his country was a good one.
He said it was; he liked it a good deal better than
any other. I asked him why. Because, he said, it
had a great many rats. I asked him if they ever
quarreled about their rat country. He said they

did. So it would appear that civilized nations are not the only people who go to war about their domain."

The writer closes this account of the Great Basin of Utah, with the following correspondence on some of the subjects to which it relates:

"WASHINGTON, June 14, 1860.

" PROFESSOR ARNOLD GUYOT, LL.D.,

"*Princeton, New Jersey.*

"DEAR SIR,—Permit me to bring to your knowledge that I have, in my explorations across the continent, given to a very conspicuous range of mountains over which I passed, the name of your *worthy self,* by which I feel that I honor less a distinguished votary of physical science than I do honor to myself. Surely one who has spoken so modestly, so adoringly, so well of Nature as the handiwork of the great *I Am*, and has shown that she and history are but the counterparts of each other, both illustrating and de-.veloping the Great Intelligent First Cause, and His goodness in thus 'arranging all things for the education of man and the realization of the plans of *His* mercy,' deserves this small tribute of respect and praise; and I bestow it, as I have said, feeling that I do not more honor a great physicist than I honor myself.

"The range of mountains, to which on my forthcoming map I have given the name of *Guyot Range,**

---

* Captain Simpson has in every instance preserved the Indian names of the ranges of mountains, where he could learn them.

is a very conspicuous one, trending north and south, and stretching from the southern shore of Great Salt Lake well on toward the Sevier River. It lies about thirty-five miles west of the valley of the Jordan and of Lake Utah. The pass through it, which my routes to California from Camp Floyd take, is a fine one, and I have, with his permission, called it after General A. S. Johnston, the distinguished officer of the army who has recently been in the command of the forces in Utah. Its altitude above the sea is six thousand two hundred and twenty-seven feet. That of the highest peaks of the range is probably about two thousand feet higher.

" My map, profiles, and report are nearly finished, but not sufficiently so to be presented to Congress for publication at its *present session.*

" I inclose a paper read before the Academy of Natural Sciences of Philadelphia, anticipatory of my more elaborate report, in reference to the *palæonto-logical* collections of my expedition. This may soon be followed up with a publication, by the same society, of some extracts from my report, which will be more particularly descriptive of the new species of fossils which were found.

" My report, I think, among other things, will illustrate in the low type of man to be found in the Indians of the 'Great Basin' of our continent, called 'Root Diggers,' *how intimately connected with the contour, re-lief, and relative position of the crust of the earth, is the development of the human race;* and will add one more to the many proofs which you have given, in your

'Earth and Man,' of this important geographical truth.

"Permit me to subscribe myself,

"Very respectfully and truly, yours,

"J. H. SIMPSON,

"Captain Topographical Engineers."

"PRINCETON, N. J., June 20, 1860.

"DEAR SIR,—I have the honor to acknowledge the receipt of your most acceptable letter, and I thank you very heartily for the kind feelings expressed in it. Guyot Range of mountains will recall to my mind more than a lofty mountain chain; it will tell me of the sympathy that truths dear to me, because fruitful of much good and enjoyment for me and for many others, have found with you. Believe me, dear sir, when I say that I feel particularly gratified to find a man of your busy profession and of your attainments so well acquainted with and so appreciative of the views too briefly exposed in the little volume to which you allude in so kind terms. Common convictions and a common faith on such grand topics are a bond of union among men which cannot easily be broken. So I shall now feel when thinking of you.

"I have read with great interest the geological notice of Messrs. Meek and Engelmann on your geological discoveries. The presence of all the great geological formations, from the Silurian and Devonian up to the Tertiary, in the Great Basin, and also the circumstance of the palæozoic rocks constituting the chief formations west of the Salt Lake, are data which throw much light on the geological history of this continent.

"I shall look with eagerness to your coming report for more light still on these regions so long unknown, and I am very glad that you did not forget the study of the poor human beings who were the first tenants of these wildernesses, and of the influence that the niggard nature, amid which their lot is cast, had in shaping their present condition.

"I remain, my dear sir, with great regard, and very truly, yours,

"A. GUYOT.

"To CAPTAIN J. H. SIMPSON,
        "*Topographical Engineers, U. S. A.*"

FINIS.

www.ingramcontent.com/pod-product-compliance
Lightning Source LLC
Chambersburg PA
CBHW031755090426
42739CB00008B/1016